This is a book about mathematical logic. It shows that a king's problems about allowing friendly neighbors into his castle, and keeping unfriendly neighbors out, are very like problems engineers face with train switches, and problems mathematicians, logicians, and ordinary people face with ordinary statements.

The king's solution to his problem uses four drawbridges across the castle moat. They illustrate the difference between AND situations and OR situations.

The king's guard-in-charge-of-drawbridges, a clever fellow, develops some tables that show all possible drawbridge situations. These illustrate TRUTH TABLES.

After exploring the king's problem and surprisingly similar switching problems with train tracks, the book concludes with an enlightening section that applies the AND-OR differences and TRUTH TABLES to ordinary statements.

Intelligent, lighthearted illustrations are by Don Madden.

YES-NO ; STOP-GO

Some Patterns in Mathematical Logic

By Judith L. Gersting and Joseph E. Kuczkowski

illustrated by Don Madden

Thomas Y. Crowell Company

New York

YOUNG MATH BOOKS

Edited by Dr. Max Beberman, Director of the Committee on School Mathematics Projects, University of Illinois

BIGGER AND SMALLER

CIRCLES

COMPUTERS

THE ELLIPSE

ESTIMATION

FRACTIONS ARE PARTS OF THINGS

GRAPH GAMES

LINES, SEGMENTS, POLYGONS

LONG, SHORT, HIGH, LOW, THIN, WIDE

MATHEMATICAL GAMES FOR ONE OR TWO

ODDS AND EVENS

PROBABILITY

RIGHT ANGLES: Paper-Folding Geometry

RUBBER BANDS, BASEBALLS AND DOUGHNUTS: A Book About Topology

STRAIGHT LINES, PARALLEL LINES, PERPENDICULAR LINES

WEIGHING & BALANCING

WHAT IS SYMMETRY?

Edited by Dorothy Bloomfield, Mathematics Specialist, Bank Street College of Education

ANGLES ARE EASY AS PIE

AREA

AVERAGES

BASE FIVE

BINARY NUMBERS

BUILDING TABLES ON TABLES: A Book About Multiplication

EXPLORING TRIANGLES: Paper-Folding Geometry

A GAME OF FUNCTIONS

HOW DID NUMBERS BEGIN?

HOW LITTLE AND HOW MUCH: A Book About Scales

LESS THAN NOTHING IS REALLY SOMETHING

MAPS, TRACKS, AND THE BRIDGES OF KÖNIGSBERG: A Book About Networks

MEASURE WITH METRIC

NUMBER IDEAS THROUGH PICTURES

ROMAN NUMERALS

SHADOW GEOMETRY

666 JELLYBEANS! ALL THAT? An Introduction to Algebra

SOLOMON GRUNDY, BORN ON ONEDAY: A Finite Arithmetic Puzzle

SPIRALS

STATISTICS

3D, 2D, 1D

VENN DIAGRAMS

YES-NO; STOP-GO: Some Patterns in Logic

 Published simultaneously in Canada by Fitzhenry & Whiteside Limited, Toronto. Manufactured in the United States of America.

Library of Congress Cataloging in Publication Data. Gersting, Judith L. Yes—no, stop—go. SUMMARY: A brief introduction to switching patterns in mathematical logic. 1. Logic, Symbolic and mathematical—Juv. lit. 2. Switching theory—Juv. lit. [1. Mathematics] I. Kuczkowski, Joseph E., joint author. II. Madden, Don. III. Title.
QA9.G36 511′.3 76-46376 ISBN 0-690-01130-X (CQR).

1 2 3 4 5 6 7 8 9 10

YOUNG MATH BOOKS

Did you ever notice that lots of things around us only come in two ways?

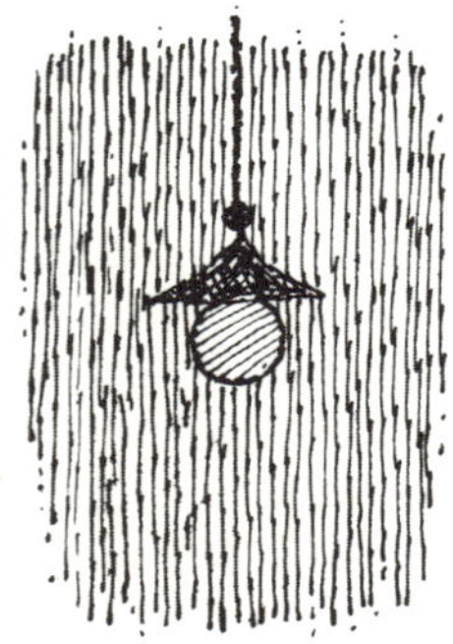

A light is on or off.

A baby is asleep or awake.

A door is open or closed.

A drawbridge is up or down.

We have a story about drawbridges.

Once upon a time there was a castle surrounded by a moat. On one side of the castle lived the friendly people of Whiffle.

On the other side of the castle lived the unfriendly people of Piffle.

The king who lived in the castle wanted to build some drawbridges across the moat to protect the castle.

On the side of the castle that faced the friendly land of Whiffle, the king's engineers built two bridges.

The bridges looked like this:

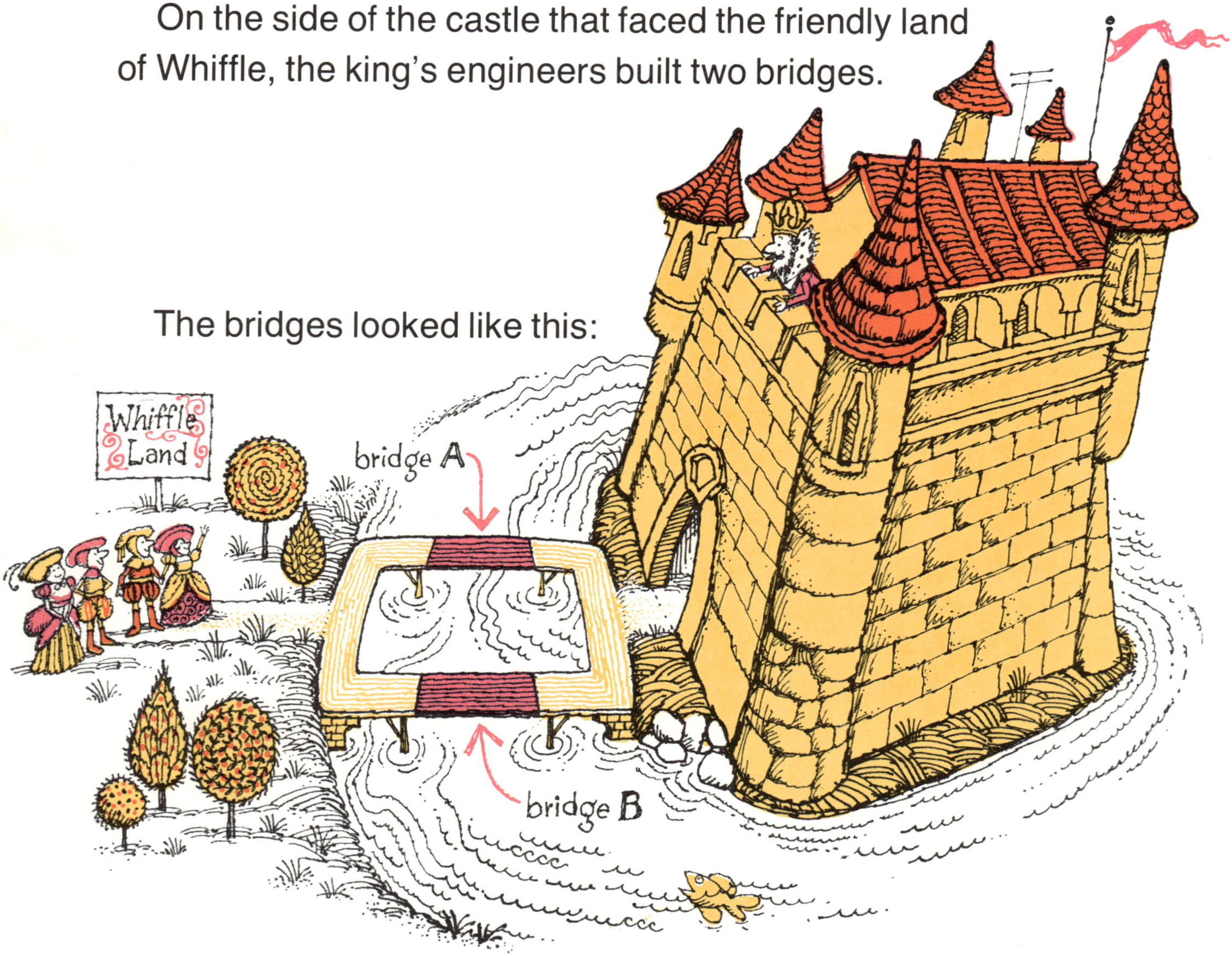

The engineers made it easy for the friendly Whifflers to get to the castle. The Whifflers could cross over to the castle whenever bridge A was down OR bridge B was down. They could certainly cross over when both bridges were down.

The engineers called this an OR pattern of bridges.

Sometimes the king did not want any visitors, so he asked that both of the bridges be pulled up. When both the bridges were up, no Whiffler could reach the castle.

But on most days the cry was heard, "Goody! At least one of the bridges is down and I can whiffle on through to the castle."

In the top turret of the castle lived the guard who ran the drawbridges. The guard wanted to have a table to tell him when a Whiffler could get to the castle on the OR pattern of bridges. He began to make his table by drawing three columns running up and down, one named A down, one named B down, and one named Whifflers can cross.

One day the guard watched the bridges very closely; he saw there were four different ways the two bridges could work together.

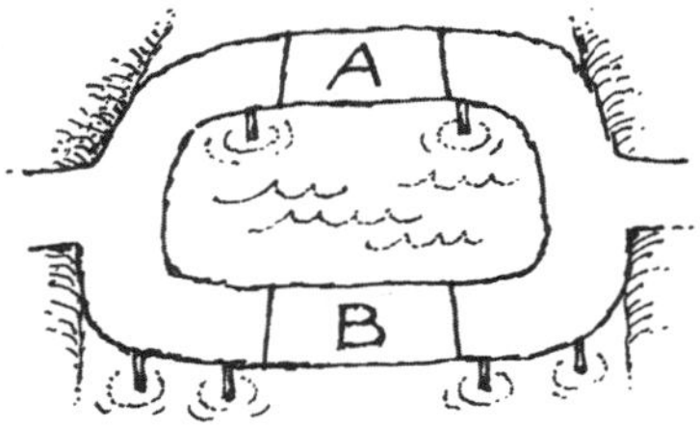

bridge A down
bridge B down

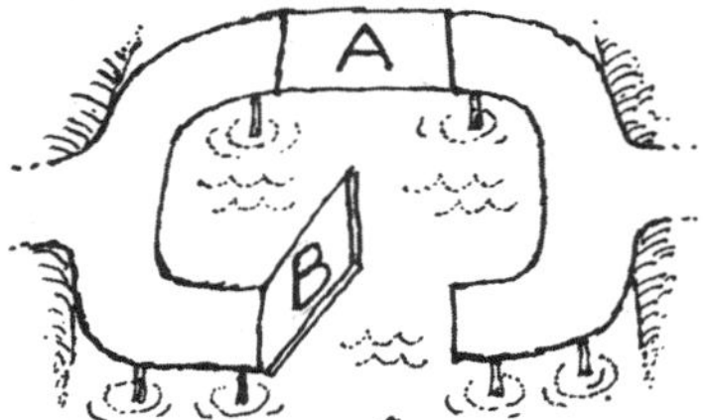

bridge A down
bridge B up

bridge A up
bridge B down

bridge A up
bridge B up

To show when a bridge was down and could be crossed, the guard wrote the word *yes* in the column naming that bridge. When a bridge was up and could not be crossed, he wrote the word *no* in the column naming that bridge.

This is how the guard began to fill out his table. Can you copy the table and finish filling in the first two columns?

Each row of the table told the guard how the two bridges looked. Knowing how the two bridges looked, the guard asked himself whether or not a Whiffler could get to the castle on the OR pattern of bridges. In the column called Whifflers can cross on his table he put the word *yes* if the answer was "Yes" and he put the word *no* if the answer was "No."

The guard knew that when both bridges were down, the Whifflers could get to the castle on the OR pattern of bridges. So, the guard put the word *yes* in the first row of his table in the column named Whifflers can cross.

A OR B pattern

A down	B down	Whifflers can cross
yes	yes	yes
yes	no	
no	yes	
no	no	

Here is how the guard's OR table finally looked:

A OR B pattern

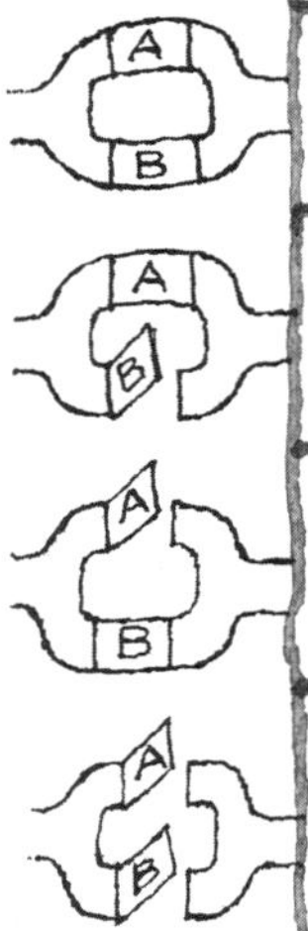

A down	B down	Whifflers can cross
yes	yes	yes
yes	no	yes
no	yes	yes
no	no	no

The guard did his work well, because his table says that the only time a Whiffler could *not* get to the castle on the OR pattern of bridges was when both bridges were up. And that's right.

On the side of the castle facing the unfriendly land of Piffle, the king's engineers also built two bridges.

Here is a picture of the bridges:

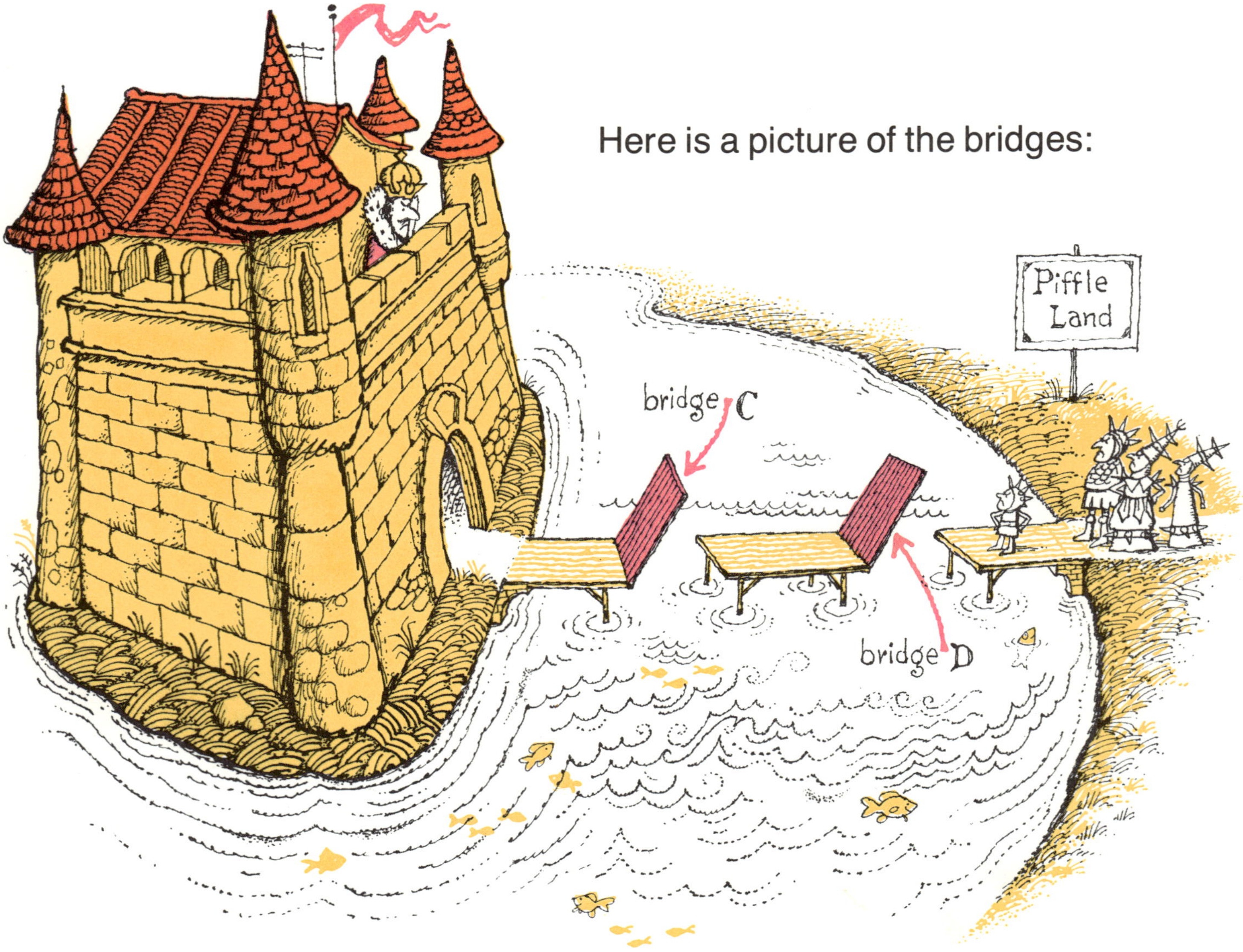

Oh, those clever engineers.

Both bridge C AND bridge D had to be down in order for a Piffler to make his way to the castle.

The engineers called this an AND pattern of bridges.

On this side of the castle the cry was often heard, "Oh, piffle! Only one of the bridges is down and I can't get through to the castle."

The guard also made a table for the AND pattern of bridges. He used the words *yes* and *no* just as before; a *yes* meant that a bridge was down and a *no* meant that a bridge was up. And, just as before, he saw that there were four different ways for bridges C and D to work together.

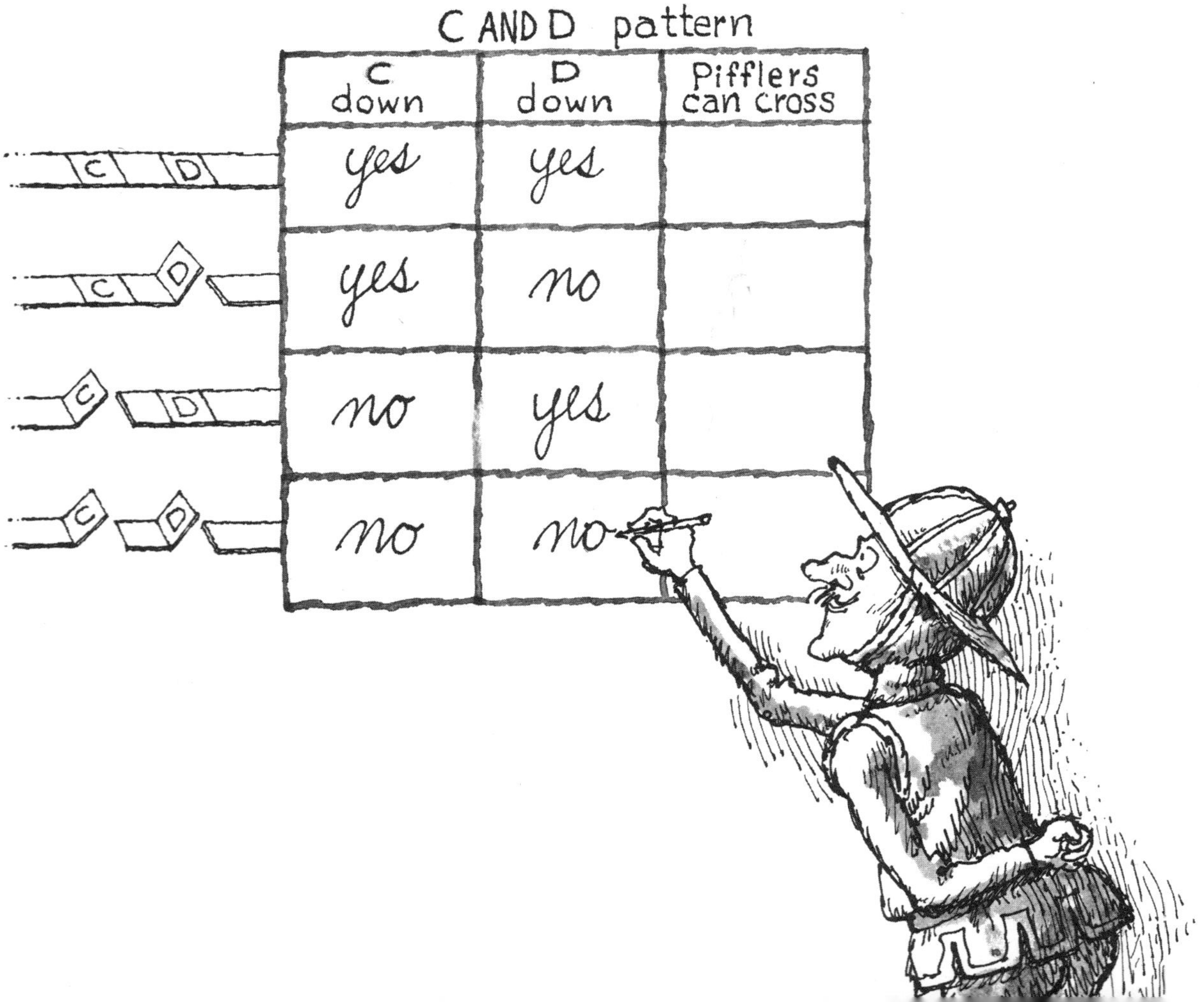

For each row of the AND table, the guard asked himself whether or not a Piffler could get to the castle on the AND pattern of bridges. In the first row, with both bridges down, the answer was *yes.* But in the second row, with bridge D up, the answer was *no.*

When the guard was halfway through with his AND table, this is how it looked. Copy the table and finish the rest of it.

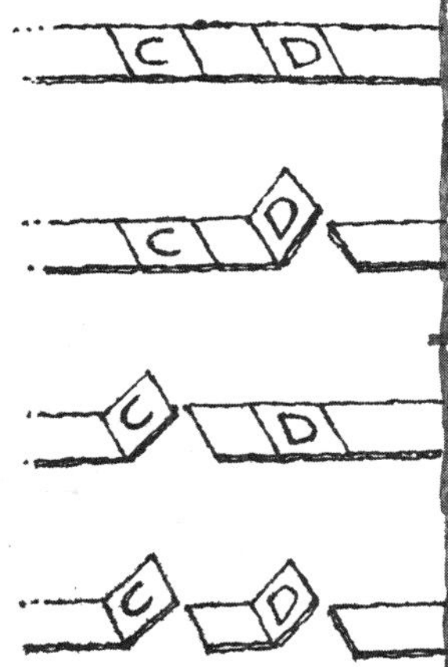

C AND D pattern

C down	D down	Pifflers can cross
yes	yes	yes
yes	no	no
no	yes	
no	no	

Did you and the guard put *no* in both empty spaces in the last column? Good. A Piffler can only get through the AND pattern of bridges when both bridges C AND D are down.

The king was so happy with the guard's work that the guard was made a knight and lived happily ever after.

Although our story was only a fairy tale, tables just like those the castle guard made have many uses today. There are lots of things around us which look like OR patterns or AND patterns.

Let's follow some train engines into a switching yard. In lots of places along the tracks in and out of the yard there are switches like this one:

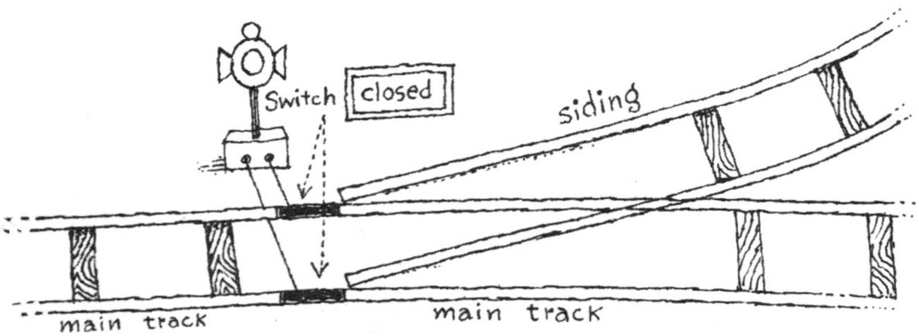

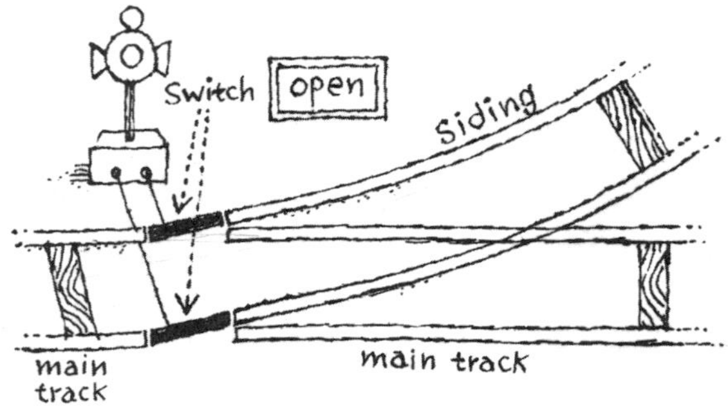

When the switch is closed a train can get through the switch and go straight along the main track.

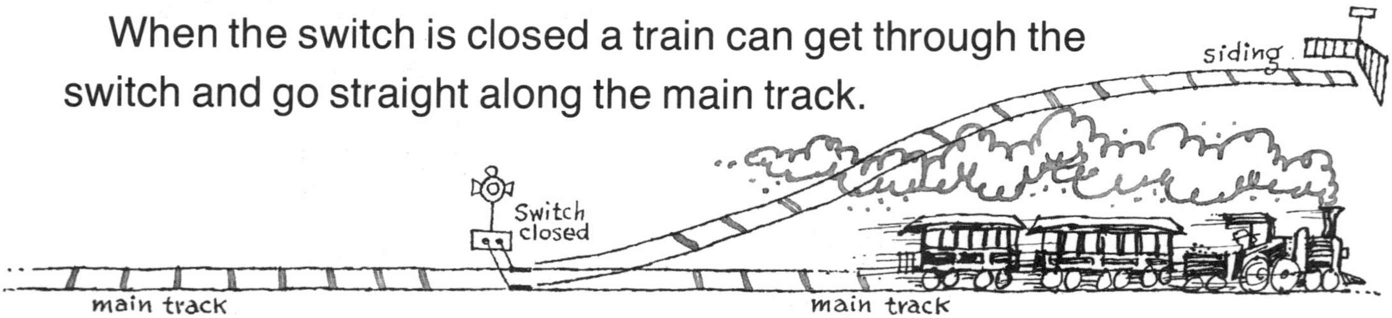

But when the switch is open the train cannot get through, so it must leave the main track and stop.

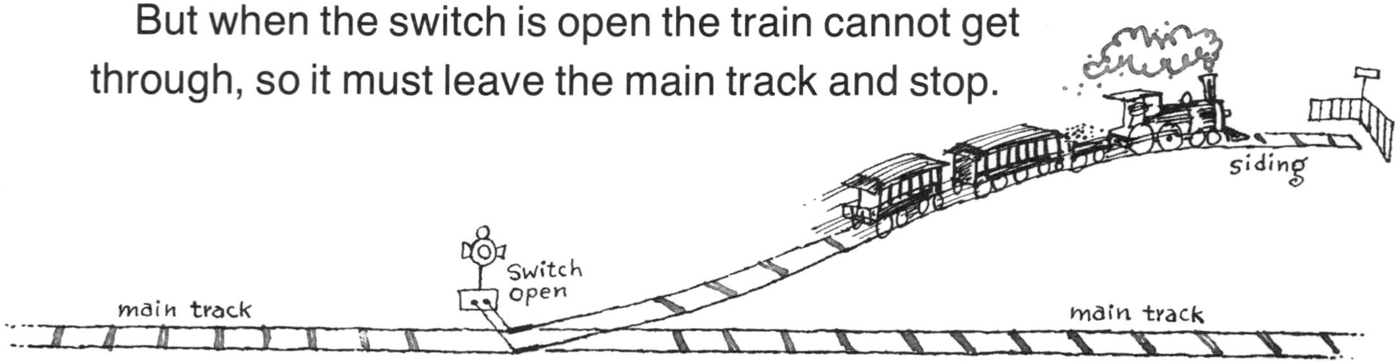

You can build models of the train switches using some large pieces of thick cardboard, some Popsicle sticks, paste, and thumbtacks.

To build a model of a single switch, paste down on a thick piece of cardboard the Popsicle sticks that are part of the main track and leave a space for the switch. Use one stick for the switch but do not paste it down. Hold it down with a thumbtack through one end of the stick and into the cardboard. You should now be able to swing the stick open and closed. Now add Popsicle track for a siding.

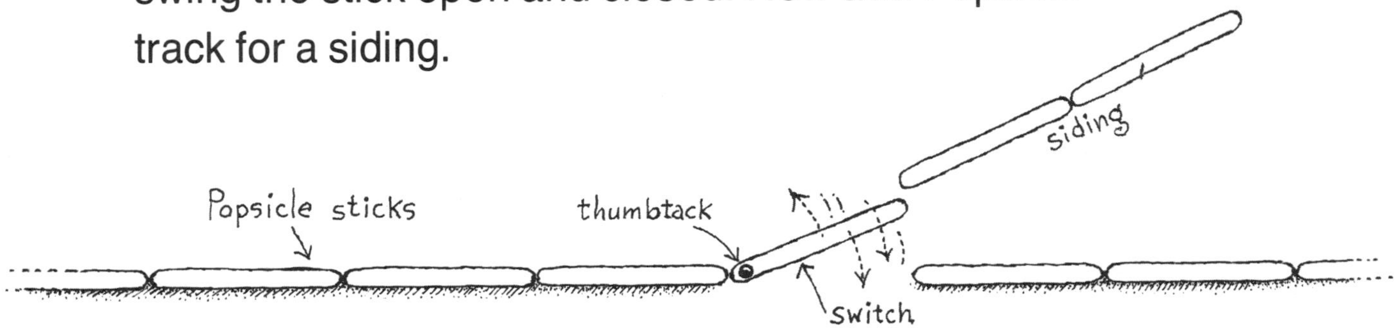

Using your cardboard and Popsicle sticks, make a model of a pattern of two train switches like this one:

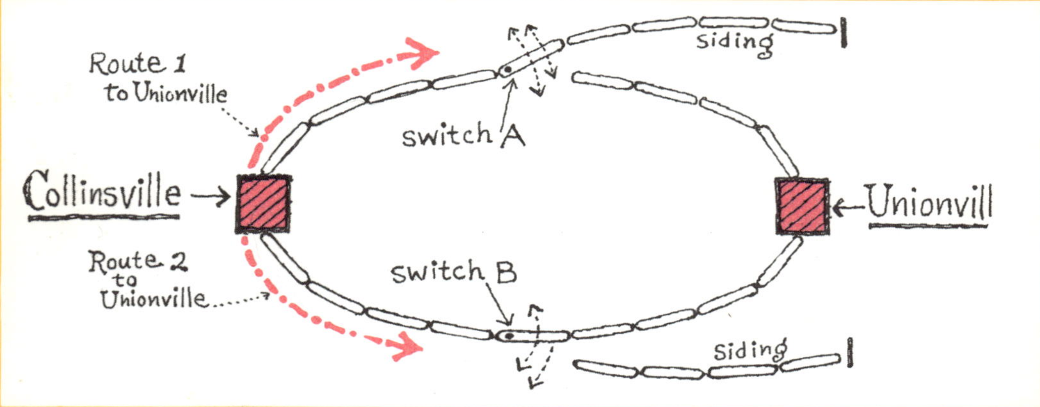

Now use your model and pretend your finger is a train traveling along the track from Collinsville to Unionville. Can a train get through when only switch A is closed? Can a train get through when only switch B is closed? What happens when both switches are closed? When is the only time a train cannot get through?

These switches work very much like the OR pattern of bridges.

You can make an OR table to show when a train can go through this pattern of switches. First draw three columns, one named switch A, one named switch B, and one named Train.

To show when a train can go through a switch, use the word *go.* To show when a train cannot get through a switch and must stop, use the word *stop*. Now fill in the first two columns so that your table looks like this:

A OR B pattern

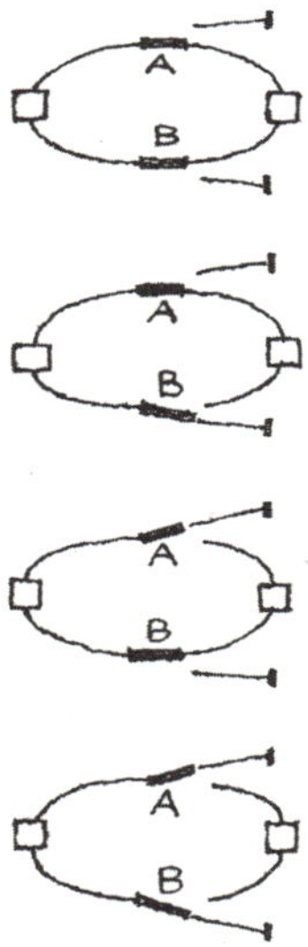

switch A	switch B	Train
go	go	
go	stop	
stop	go	
stop	stop	

In the first row of the table, both switch A and switch B are closed (marked *go*). When both switches are closed, you know that a train can get through, so you should put the word *go* in the top box of the last column. Now finish filling in your table for the OR pattern of train switches.

Your table for the OR pattern of switches should look like this:

A OR B pattern

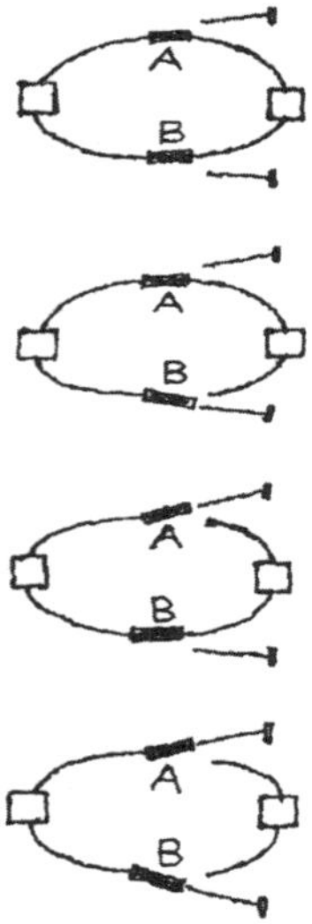

switch A	switch B	Train
go	go	go
go	stop	go
stop	go	go
stop	stop	stop

Take your table and put it next to the table for the OR pattern of bridges (page 11). The words are different, but otherwise the two patterns are just alike.

You can also build a model of another pattern of switches often found in the switching yard.

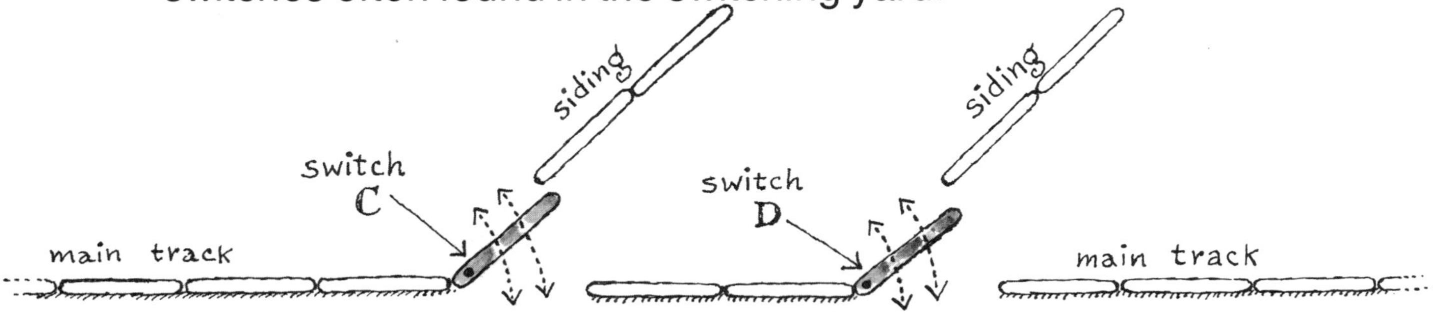

What switches must be closed here for a train to be able to go along the main track?

Doesn't this pattern of switches make you think of the AND pattern of bridges?

Can you copy and finish the table for the AND pattern of switches?

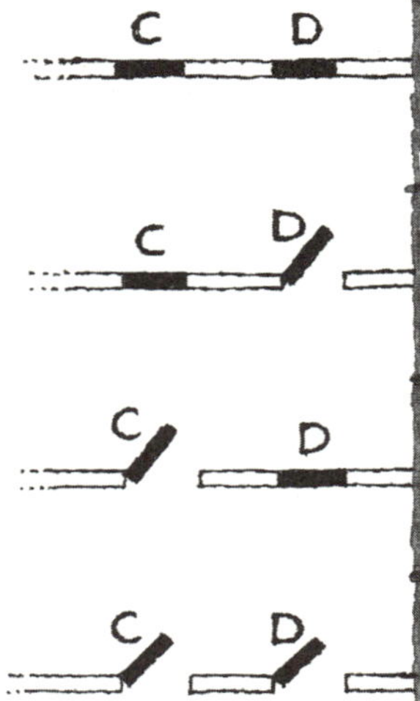

C AND D pattern

switch **C**	switch **D**	Train
go	go	
go	stop	stop
stop	go	
stop	stop	stop

Your table should look like the one for the AND pattern of bridges, but with different words of course.

OR and AND patterns also appear in a part of mathematics called *logic.* Just as bridges can be down or up and switches closed or open, certain sentences, called *statements,* are true or false.

"George Washington was the first president of the United States" is a true statement. "The moon is made of green cheese" is a false statement.

Some sentences, such as "Stop!", are not statements because they are not true or false. Questions are also not statements because they are not true or false.

Look on page 16 of this book; can you find a sentence which is a statement? Can you find a sentence which is not a statement?

Think up some statements of your own. Write them down on a piece of paper. Make sure there are both true and false statements on your list. Mark the true statements T and the false statements F.

Here is a beginning:

If we have two statements we can form a new statement by putting them together with the word OR. Let's take two true statements. "George Washington was the first president of the United States or Halloween is in October." This new statement is also true.

Put two of your own true statements together with OR. Is the new statement true?

Now try putting two false statements together with the word OR. Try putting together a true statement and a false statement; try a false and a true statement. Which of the new statements are true? Which are false?

When two statements are put together with the word OR, the new statement is true when either of the old statements is true or when both of the old statements are true.

Try making a table for statements put together with OR. You can use the word *true* to show when a statement is true and the word *false* to show when a statement is false.

A table like this is called a truth table. This truth table looks very much like all our other OR tables.

We can also put two statements together with the word AND. Choose two true statements from your list and put them together with the word AND. Is the new statement true?

Is the statement "George Washington was the first president of the United States and the moon is made of green cheese" a false statement?

Make an AND truth table. Put two true statements together with AND. Next put a true statement and a false statement together. Then try a false and a true statement. Last put two false statements together with the word AND. Now fill in the last column of your table.

Does this table remind you of any others you have seen?

Truth table

	statement A	statement B	statement A AND B
Dogs like bones and cats like fish.	true	true	
Dogs bark and cats have wings.	true	false	
Dogs say meow and cats purr.	false	true	
Dogs read books and cats write letters.	false	false	

Can you think of other places where you have met OR patterns?

Can you think of other places where you have met AND patterns?

About the Authors

Judith L. Gersting and Joseph E. Kuczkowski both teach mathematics at Indiana University–Purdue University at Indianapolis. Together they have written several mathematics books for children. YES-NO; STOP-GO, a book about logic, is their first Crowell Young Math Book.

Dr. Gersting lives in Indianapolis with her husband and two children, Adam and Jason. She is especially interested in mathematical logic, and likes to cook and do needlework in her spare time.

Dr. Kuczkowski also lives in Indianapolis with his wife and three children, Edward, Ann, and Jim. His hobbies include basketball, singing, and playing the piano.

About the Illustrator

Don Madden attended the Philadelphia Museum College of Art on a full scholarship. Following graduation, he joined the faculty as an instructor in experimental drawing and design. The recipient of gold and silver medals at the Philadelphia Art Directors' Club exhibitions, Mr. Madden has had his work selected for reproduction in the *New York Art Directors' Annual,* in the international advertising art publication *Graphis,* and in the *Society of Illustrators Annual.*

Don Madden lives and works in a restored old farmhouse in upstate New York. He has a wife, who is also an artist, two children, and a flock of geese to keep him company.